JWST FROM BELOW

RAYMOND OHL

NIMBLE BOOKS LLC: THE AI LAB FOR BOOK-LOVERS
~ FRED ZIMMERMAN, EDITOR ~

Humans and AI making books richer, more diverse, and more surprising.

Publishing Information

ISBN: 978-1-60888-286-1

AI-generated Keyword Phrases

James Webb Space Telescope; NASA; Optics Branch; NASA Goddard Space Flight Center; telescope development; science themes; early universe; galaxy formation and evolution; star formation in our galaxy; planetary systems; telescopes for observing distant objects; space as preferred environment for telescopes; Optical Telescope Element (OTE); Integrated Science Instrument Module (ISIM); construction of JWST; testing of JWST; deployment of JWST; launch of JWST; additional resources on JWST.

Publisher's Notes

This document provides an overview of the construction and mission of the James Webb Space Telescope (JWST), which is set to revolutionize our understanding of the universe. Readers should want to read this document because it highlights the cutting-edge technology and scientific advancements that JWST represents, offering insights into the mysteries of the early universe, galaxy formation, star formation, and planetary systems. By exploring these topics, the document connects with current public issues such as the search for extraterrestrial life, the origins of the universe, and the impact of space exploration on our understanding of Earth and our place in the cosmos.

This annotated edition illustrates the capabilities of the AI Lab for Book-Lovers to add context and ease-of-use to manuscripts. It includes several types of abstracts, building from simplest to more complex: TLDR (one word), ELI5, TLDR (vanilla), Scientific Style, and Action Items; essays to increase viewpoint diversity, such as Grounds for Dissent, Red Team Critique, and MAGA

Perspective; and Notable Passages and Nutshell Summaries for each page.

ANNOTATIONS

ABSTRACTS

TL;DR (ONE WORD)

JWST.

EXPLAIN IT TO ME LIKE I'M FIVE YEARS OLD

The James Webb Space Telescope (JWST) is a really big telescope that NASA is building to study things far away in space. It's like a giant eye in the sky that can see things that our regular eyes can't.

NASA and a special group at NASA called the Optics Branch are working together to make this telescope. They are making sure it's built correctly and will work well in space.

The JWST has four important things it wants to study. First, it wants

TL;DR (VANILLA)

This document provides an overview of the James Webb Space Telescope (JWST), discussing its construction, mission, and science themes. It explains why telescopes are used to observe distant objects and why space is preferred for telescopes. It also provides information on the different elements of JWST and its construction, testing, deployment, and launch. Additional resources for more information are included.

SCIENTIFIC STYLE

This document provides a comprehensive overview of the construction and mission of the James Webb Space Telescope (JWST). It discusses the involvement of NASA and the Optics Branch at NASA Goddard Space Flight Center in the development of JWST. The document highlights the four science themes of JWST, which focus on studying various aspects of the universe. It also explains why telescopes are used to observe distant objects and why space is the preferred environment for telescopes. Information on different elements and regions of JWST, such as OTE and ISIM, is provided. The construction, testing, deployment, and launch processes of JWST are detailed. Links to additional resources for further information on JWST are included.

ACTION ITEMS

Read the document to gain a comprehensive understanding of the construction and mission of the James Webb Space Telescope.

Explore the four science themes of JWST and their significance in advancing our knowledge of the universe.

Learn about the role of NASA and the Optics Branch at NASA Goddard Space Flight Center in developing JWST.

Understand why telescopes are used to observe distant objects and why space is preferred for telescopes.

Familiarize yourself with the different elements and regions of JWST, such as OTE and ISIM.

Study the construction, testing, deployment, and launch processes.

VIEWPOINTS

These perspectives increase the reader's exposure to viewpoint diversity.

GROUNDS FOR DISSENT

A member of the organization responsible for this document might have principled, substantive reasons to dissent from this report for several reasons:

Misallocation of resources: The member may believe that the resources allocated for the construction and mission of the James Webb Space Telescope could be better utilized in other areas of scientific research or addressing more pressing societal needs. They may argue that investing billions of dollars into a single space telescope is not justified when there are other critical issues that require funding, such as poverty alleviation, healthcare, or climate change mitigation.

Lack of diversity in scientific exploration: The member may criticize the focus of JWST on studying distant objects and phenomena in space. They may argue that this narrow emphasis neglects other important areas of scientific exploration, such as environmental research or understanding Earth's ecosystems. By exclusively prioritizing space observation, they believe that the organization is disregarding the study of our own planet and its immediate surroundings.

Ethical concerns: A dissenting member may raise ethical concerns regarding the construction and launch of the JWST. They may argue that spending enormous amounts of resources on space exploration while many people on Earth still lack basic necessities is morally questionable. Additionally, they might question the potential environmental impact associated with constructing and launching a massive telescope into space, including waste generation and possible damage to space debris.

Scientific skepticism: The member might hold different scientific beliefs or theories that contradict some assumptions presented in the document. For example, they may dispute the notion that telescopes are necessary for observing distant objects or challenge the preference for using space as a telescope's environment. They might propose alternative methods or technologies that could achieve similar scientific goals more effectively or efficiently.

Lack of transparency or accountability: The dissenting member might believe that important information or perspectives were omitted from this document, leading to an incomplete or biased portrayal of the JWST project. They may argue that certain risks, limitations, or controversies related to the telescope's development and mission were deliberately left out, potentially misleading stakeholders or the general public.

It is important to note that these dissenting views are hypothetical and do not reflect actual criticisms of the James Webb Space Telescope. Each viewpoint is subjective and based on individual beliefs or values, which can vary within an organization.

RED TEAM CRITIQUE

Overall, the document provides a comprehensive and informative overview of the James Webb Space Telescope (JWST), its construction, and mission. However, there are several areas where improvements can be made to enhance clarity and provide additional relevant information.

Introduction: The document should include a brief introduction that captures the reader's attention and provides context for the importance of studying distant objects in space. This will help readers understand why the JWST is significant in advancing our knowledge of the universe.

Mission Objectives: While the document mentions four science themes of JWST, it could benefit from expanding on each theme with specific scientific objectives or questions being addressed. This would give readers a better understanding of how these science themes contribute to our overall understanding of astrophysics.

Technical Details: The section describing different elements and regions of JWST lacks technical details about their functioning and capabilities. Including specific information about each component's purpose within the telescope system would make this section more informative for readers interested in understanding how JWST operates.

Construction Timeline: The document briefly mentions construction and testing activities but fails to provide an overview timeline or key milestones along this process. Incorporating a timeline would allow readers to visualize progress from development to launch, providing additional context regarding project complexity.

Testing Methods: Expand on explanations of various testing methods utilized during construction phases (optical tests, vibration tests). Include details on what these tests aim to achieve, how they were conducted, and highlight any challenges encountered during this phase.

Deployment Process: Although briefly mentioned in passing, providing more details about deployment mechanisms or strategies used by JWST would enhance reader comprehension regarding logistics involved in deploying such a large-scale telescope into space successfully.

Risk Mitigation: Address potential risks associated with constructing such complex technology while highlighting actions taken by NASA or other agencies involved to mitigate those risks throughout development stages.

8.Expand Conclusion Section:

Include an expanded conclusion section that emphasizes the significance of JWST's mission and its potential impact on scientific discoveries. This could also include future expectations or goals, such as anticipated findings or specific astrophysical questions that JWST aims to answer.

Additional Resource Links: While providing links for additional information is a valuable addition, it would be beneficial to categorize these resources into sections relevant to each topic covered in the document (e.g., construction, science themes) and briefly describe their content so readers can navigate them more efficiently.

Visual Aids: Incorporate relevant images or diagrams throughout the document to assist readers in visualizing key concepts and elements discussed. For example, including an illustration of telescope deployment mechanisms or a schematic diagram depicting different components within JWST would greatly enhance clarity.

Overall, with some improvements addressing these areas of concern, the document will offer a comprehensive understanding of the construction process and the mission objectives of JWST while maintaining reader engagement.

MAGA PERSPECTIVE

This document is just another example of NASA's liberal agenda and wasteful spending. The James Webb Space Telescope (JWST) is nothing more than a pet project of the elites in the scientific community, funded by hardworking American taxpayers. Instead of focusing on issues that actually matter, like bringing jobs back to America or securing our borders, NASA is wasting money on studying distant galaxies and planetary systems.

The so-called "science themes" discussed in this document are nothing but vague concepts that serve no practical purpose for everyday Americans. What do we gain from studying the early universe or galaxy formation? These are just distractions from the real problems facing our nation.

Furthermore, why should we be investing in telescopes that observe distant objects when there are so many issues right here on Earth that need our attention? Our infrastructure is crumbling, our education system is failing, and yet NASA thinks it's more important to launch telescopes into space.

Not to mention the excessive cost of building and testing the JWST. This project has gone way over budget and has faced numerous delays. It's clear that there is mismanagement and incompetence within NASA. We can't afford to keep pouring money into projects like this when there are pressing needs that require immediate attention.

In conclusion, the JWST represents everything wrong with the priorities of our government. We need leaders who will focus on improving life here on Earth instead of indulging in expensive space exploration endeavors. It's time to put America first again.

PAGE-BY-PAGE SUMMARIES

NOTABLE PASSAGES

BODY-3   "We are living in a 'golden age' for astronomy,
          planetary science, heliophysics, Earth science and
          space science! We are on the cusp of a new golden age
          for human space flight!"

BODY-7   "To 'Drive advances in science...', in this case
          ASTROPHYSICS. Astrophysics is humankind's
          scientific endeavor to understand the universe and
          our place in it. How did our Universe begin and
          evolve? How did Galaxies, Stars, and Planets come to
          be? Are we alone?"

BODY-9   Telescopes are the natural choice of instruments used
          to directly observe distant objects that emit light.

BODY-11   "The average temperature is 2.725 Kelvin degrees
          above absolute zero (absolute zero is equivalent to -
          273.15 ºC or -459 ºF), and the colors represent the tiny
          temperature fluctuations, as in a weather map."

JWST from below*
An overview of the construction of the James Webb Space Telescope (JWST) and discussion of current status

February 8, 2023

Raymond Ohl
Branch Head, Optics Branch
NASA Goddard Space Flight Center

*Apologies to R. Feynman ("Los Alamos from Below," *Reminiscences of Los Alamos, 1943—1945*, L. Badash et al. eds., D. Reidel Publishing Co., Dordrecht, p. 105, 1980).

Outline

- **About NASA**
- **About the Optics Branch at NASA Goddard Space Flight Center (GSFC)**
- **Introduction to the James Webb Space Telescope and its mission**
- **Latest updates**

JWST construction occurred across the country and around the world (and at nearby NASA Goddard Space Flight Center, Greenbelt, Md.)

NASA

 # A little about NASA

NASA

- **US Government's civilian "space department"**
- **Partners with industry and academia, but in-house work, too**
- **Space exploration**
  - Human space flight and exploration
  - Robotic exploration
  - Astronomy and other space science
- **Earth science**
  - Weather
  - Climate change
- **Airplane research**
- **NASA works closely with companies, universities, and other countries**
- **We are living in a "golden age" for astronomy, planetary science, heliophysics, Earth science and space science!**
- **We are on the cusp of a new golden age for human space flight!**
- **Each NASA center has a different mission/purpose**
- **NASA Goddard Space Flight Center (GSFC) is focused on Earth and other science missions (not much manned space flight)**

Our Vision Statement

We reach for new heights and <u>reveal the unknown</u> for the benefit of humankind

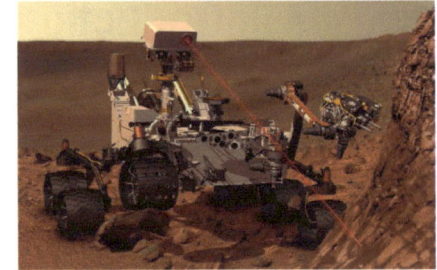

NASA

Our Mission

Drive advances in <u>science</u>, technology, aeronautics, and space exploration to enhance knowledge, education, innovation, economic vitality and stewardship of Earth

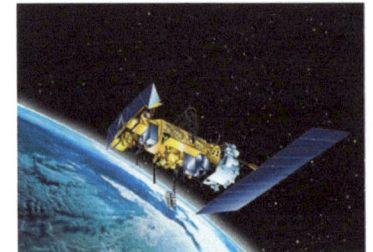

NASA

Optics Branch, GSFC

- **All phases of optical (and some non-optical) system development for all types of flight (and some non-flight) projects, including R&D and early-phase efforts**
- **Mix of civil servant and contractor personnel (~130 total)**
- **Close engagement with science and engineering groups to deliver both in-house and externally-developed telescopes and instruments**
- **Partnerships and collaborations with many elsewhere in NASA, other Government, academia, and industry**
- **Unique, world-class facilities with a broad range of capability**
- **Divided into Groups**
  - Design
  - Fabrication
  - Components
  - Alignment, Integration, and Test
  - Wavefront Sensing and Control

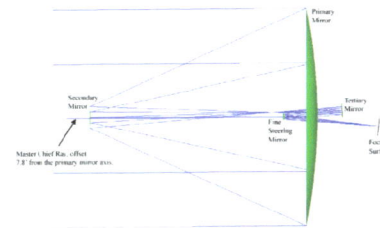

Optical layout of JWST telescope

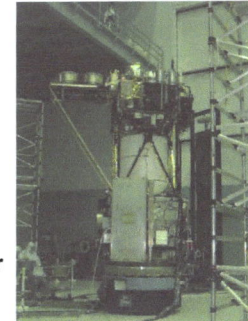

Optical metrology for the spacecraft

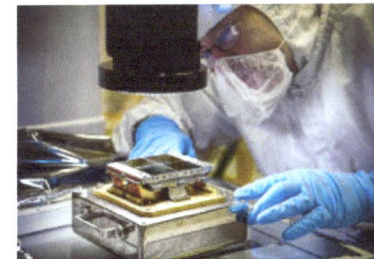

Alignment metrology for JWST NIRCam detectors

Diamond machining of optics and precision, optomechanical surfaces

OSIRIS-REx OVIRS and optics team

Image from JWST ground testing

JWST is a general astrophysics mission

- **JWST will operate in a manner similar to HST to enable a wide range of science investigations proposed by astronomers world-wide (scientific follow-on)**

- **General Observer community will drive science investigations**

- **Four science themes define the development of technical requirements for JWST:**

  - First light and reionization: Identify the first bright objects in the early Universe and follow ionization history

  - Galaxy formation and evolution: Shed light on how galaxies and dark matter evolved to present

  - Star formation in our galaxy: Study the birth and early development of stars

  - Planetary systems: Observe the physical and chemical properties of solar systems (including our own)

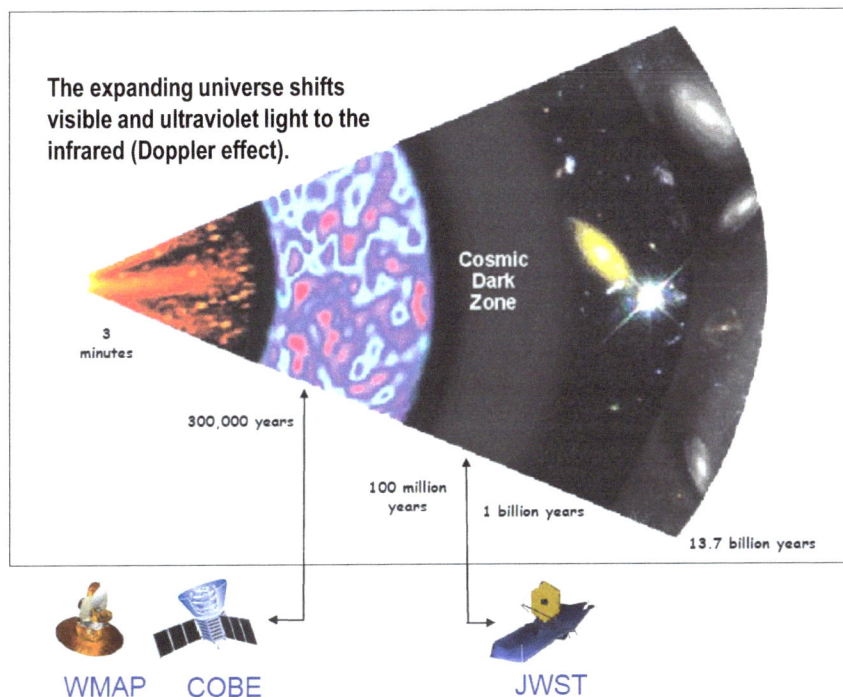

The expanding universe shifts visible and ultraviolet light to the infrared (Doppler effect).

Cosmic Dark Zone

3 minutes

300,000 years

100 million years

1 billion years

13.7 billion years

WMAP   COBE   JWST

NASA

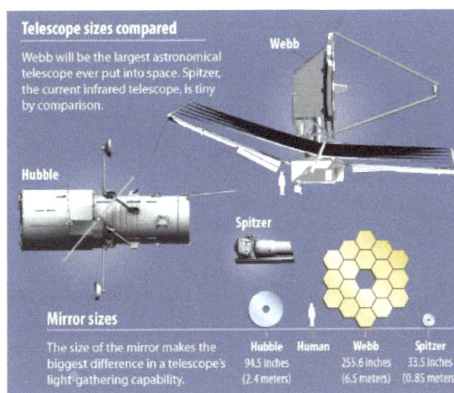

Telescope sizes compared

Webb will be the largest astronomical telescope ever put into space. Spitzer, the current infrared telescope, is tiny by comparison.

Webb

Hubble

Spitzer

Mirror sizes

The size of the mirror makes the biggest difference in a telescope's light-gathering capability.

Hubble	Human	Webb	Spitzer
94.5 inches (2.4 meters)		255.6 inches (6.5 meters)	33.5 inches (0.85 meters)

NASA

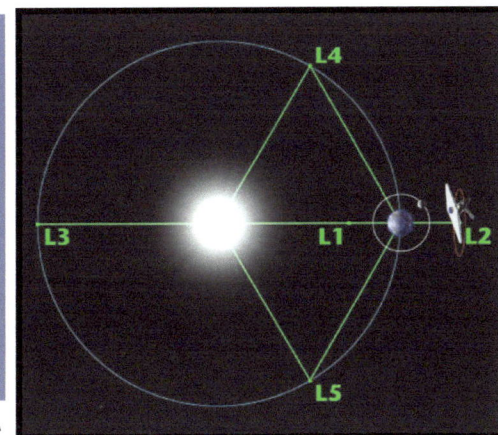

NASA

L4

L3   L1   L2

L5

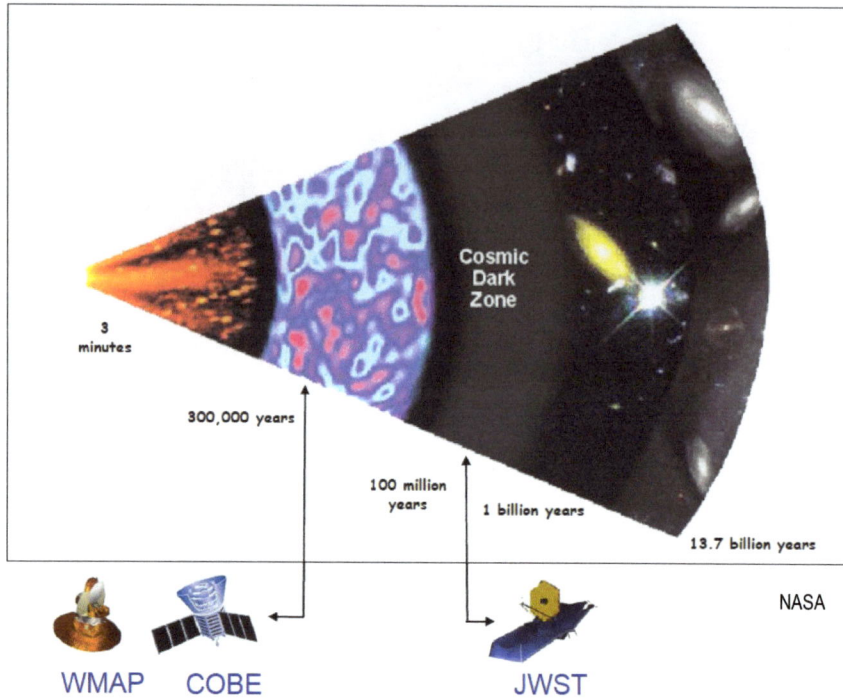

NASA

To "Drive advances in science…", in this case ASTROPHYSICS.

Astrophysics is humankind's scientific endeavor to understand the universe and our place in it.

How did our Universe begin and evolve?
How did Galaxies, Stars, and Planets come to be?
Are we alone?

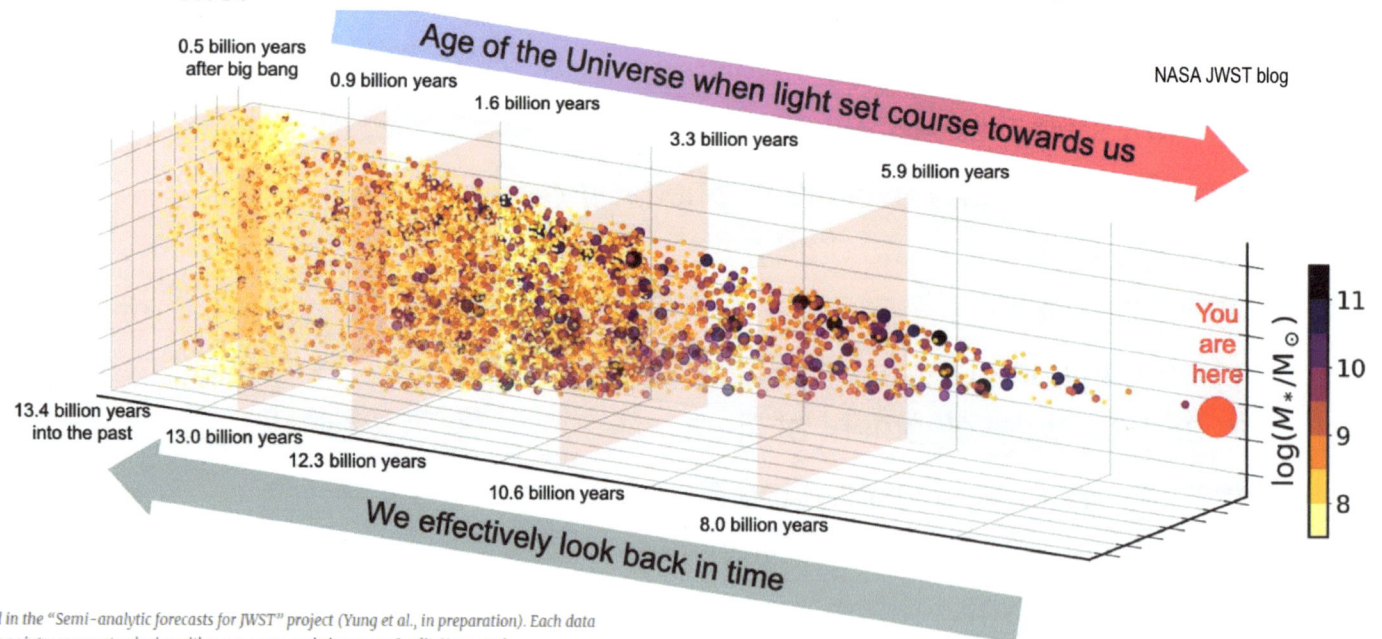

NASA JWST blog

Side-view of the simulated universe as presented in the "Semi-analytic forecasts for JWST" project (Yung et al., in preparation). Each data point represents a galaxy. Larger and darker data points represent galaxies with more mass, and vice versa. Credit: Yung et al.

How does a telescope or camera work?

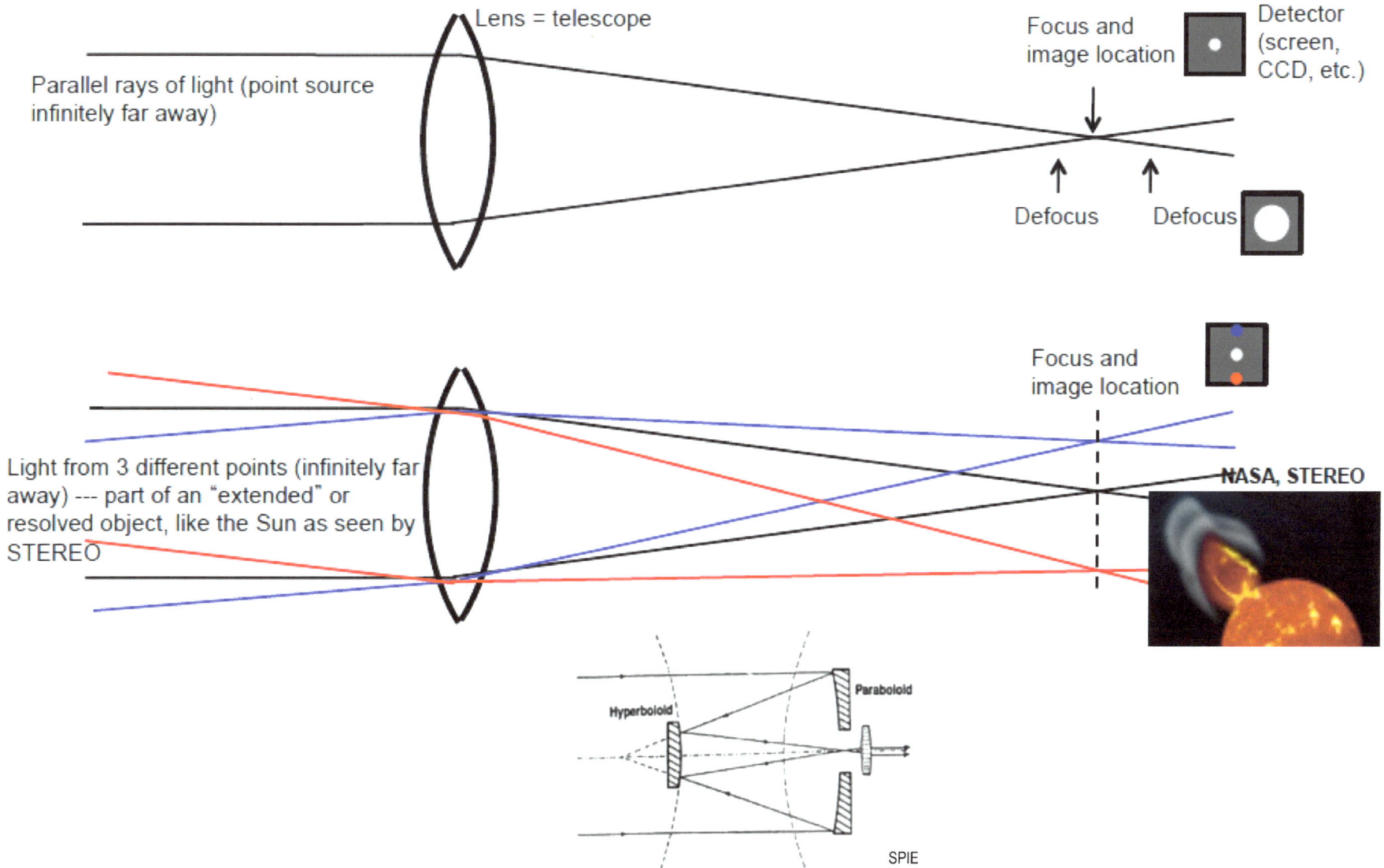

Lens = telescope

Parallel rays of light (point source infinitely far away)

Focus and image location

Detector (screen, CCD, etc.)

Defocus   Defocus

Focus and image location

Light from 3 different points (infinitely far away) --- part of an "extended" or resolved object, like the Sun as seen by STEREO

NASA, STEREO

Hyperboloid   Paraboloid

SPIE

8

Why **Telescopes**?

- **Telescopes are the natural choice of instruments used to directly observe distant objects that emit light.**

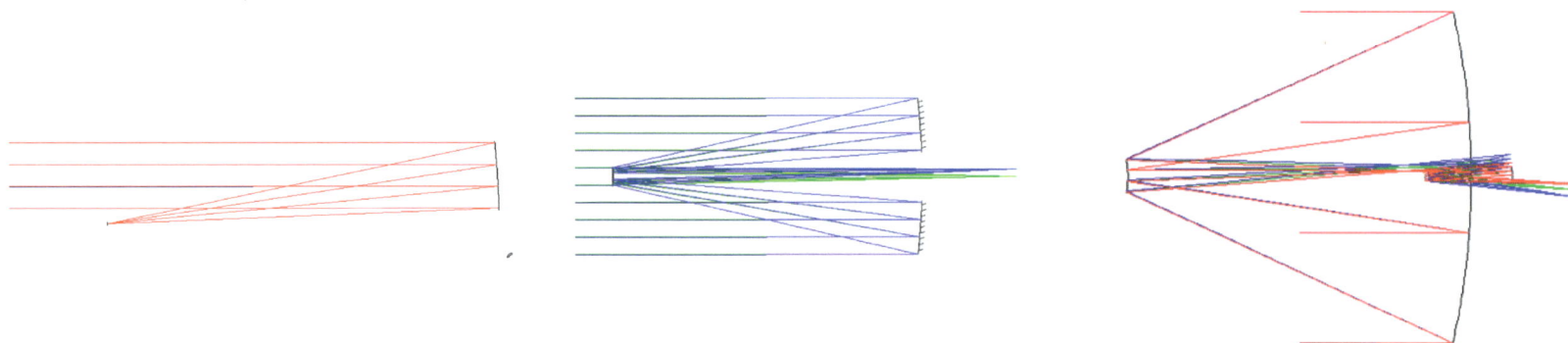

NASA

Why **Space**?

1. **No Atmosphere**
   - No turbulence = better resolution
   - No absorption = see all wavelengths
2. **Very stable environment**
   - Can orbit with constant sun exposure (no day/night thermal changes)
3. **Can get very cold**
   - Better for observing in the **Infra-Red** spectrum (old visible-UV light)

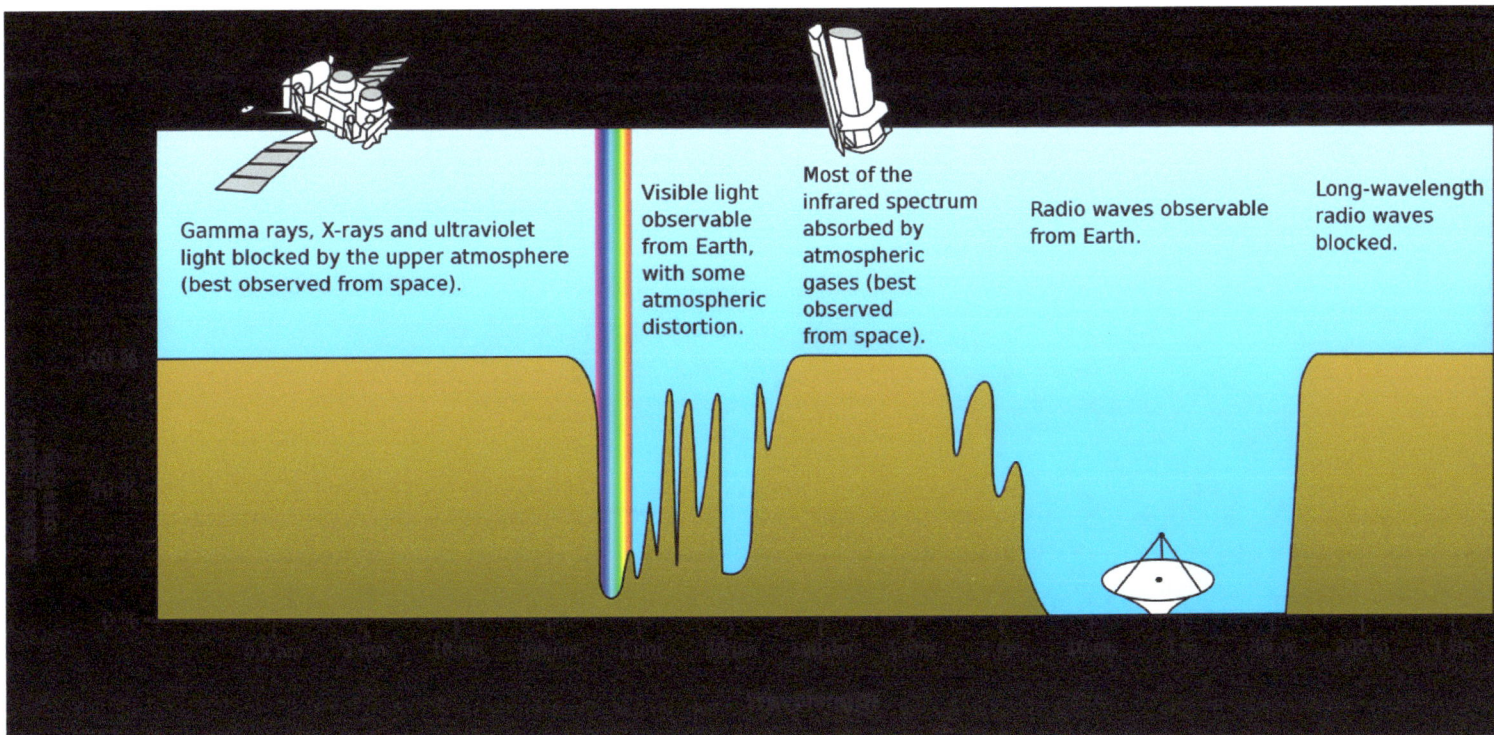

Gamma rays, X-rays and ultraviolet light blocked by the upper atmosphere (best observed from space).

Visible light observable from Earth, with some atmospheric distortion.

Most of the infrared spectrum absorbed by atmospheric gases (best observed from space).

Radio waves observable from Earth.

Long-wavelength radio waves blocked.

NASA

Why **Space**? Space background is very cold

- **Cosmic background temperature is < 3 K**
- **With sun shields and cryostats, entire observatories can be cooled actively to <10 K, or passively to <35 K**

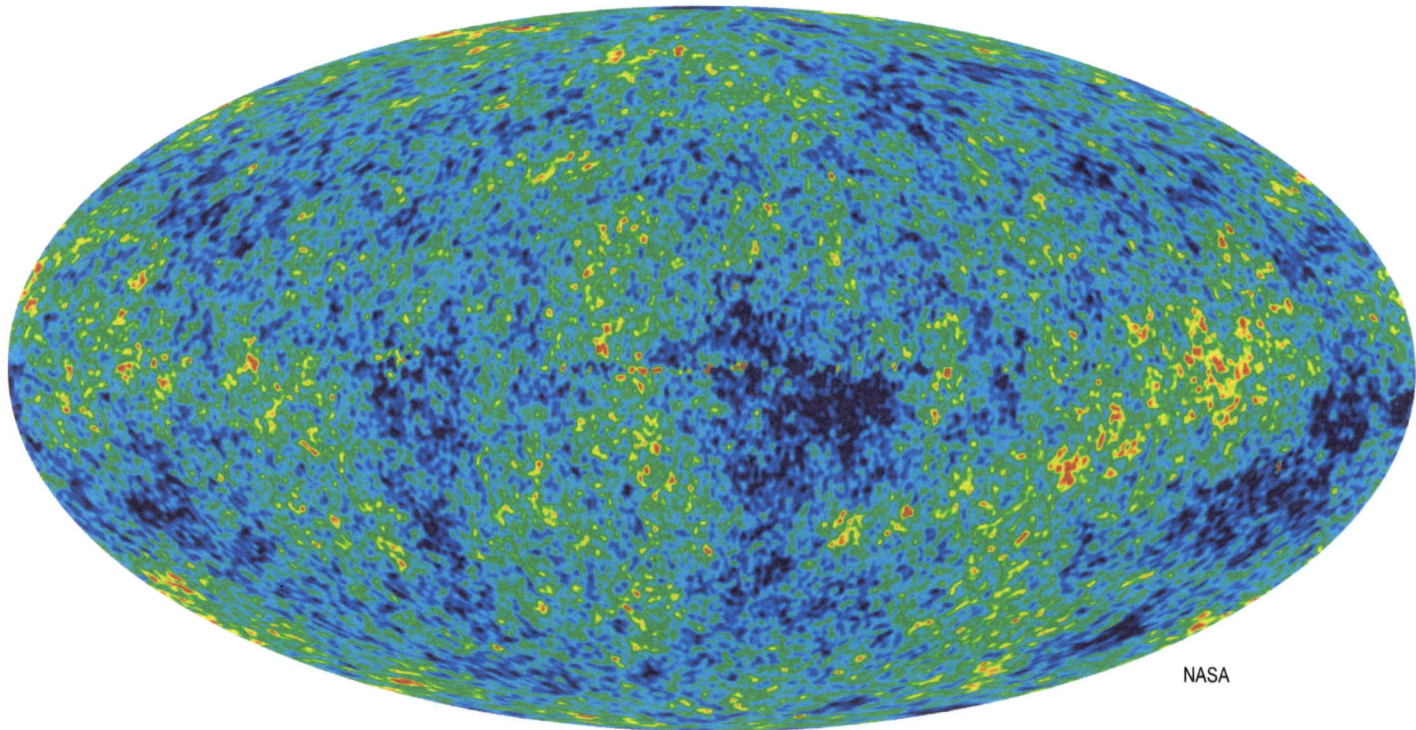

NASA

The Cosmic Microwave Background temperature fluctuations from the 7-year Wilkinson Microwave Anisotropy Probe data seen over the full sky. The image is a mollweide projection of the temperature variations over the celestial sphere. The average temperature is 2.725 Kelvin degrees above absolute zero (absolute zero is equivalent to -273.15 ºC or -459 ºF), and the colors represent the tiny temperature fluctuations, as in a weather map. Red regions are warmer and blue regions are colder by about 0.0002 degrees.

Why **Space**? Very Stable
Environment: L2

- **The Red points labeled L1 to L5 represent "stable" gravity points in the Sun-Earth system ("Lagrange" points)**
- **L1 is favored Sun-observing Space Observatories**
- **L2 is favored for some Astronomical Space Observatories**

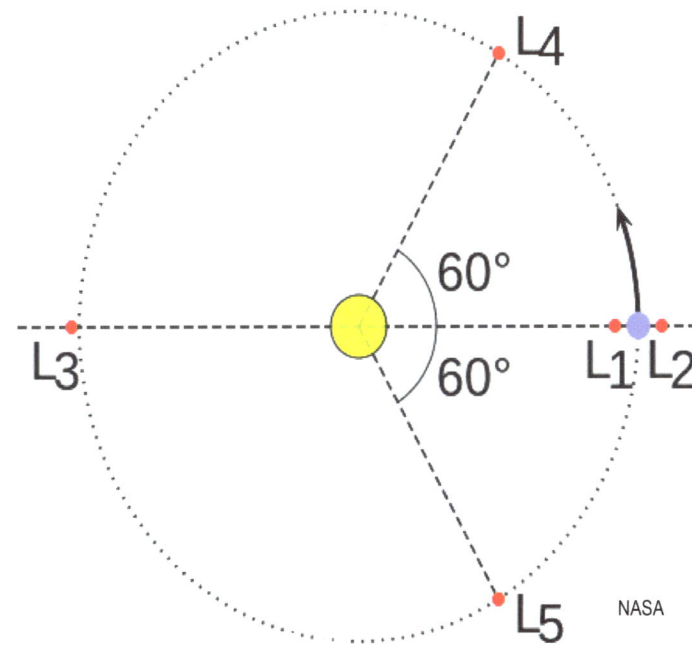

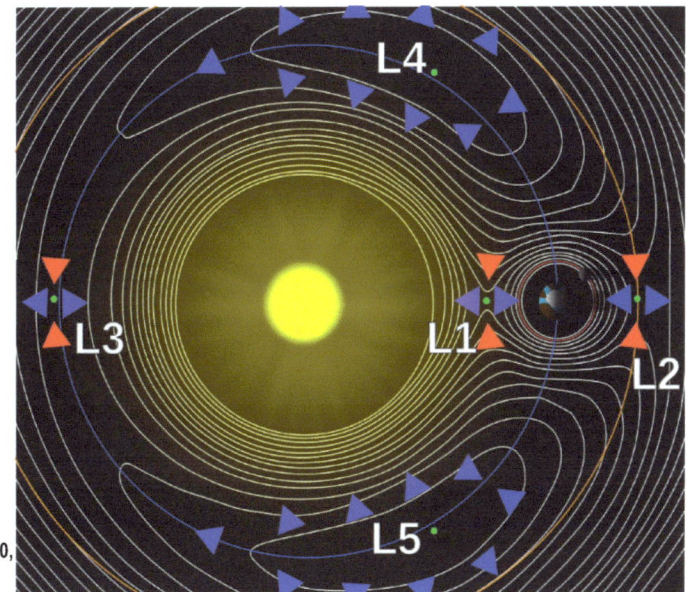

JWST must be very cold to see infrared: 30K

Recipe on how to get cold:

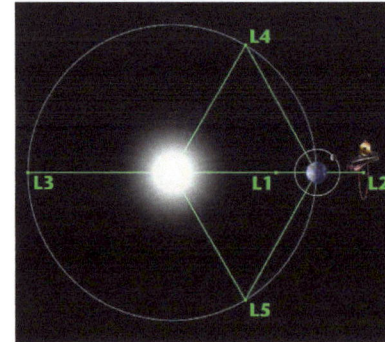

1. Get far away from Earth
   (Sun-Earth L2 point)

2. Hide in the shade

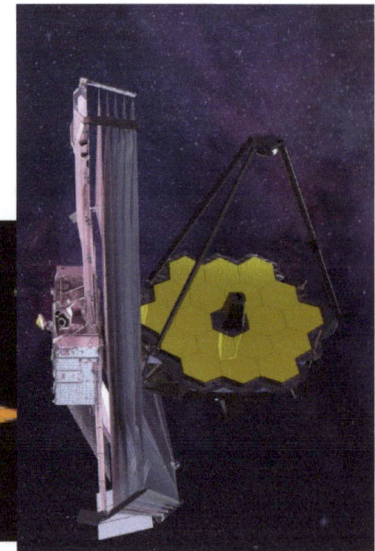

NASA

JWST Observatory Elements and Regions

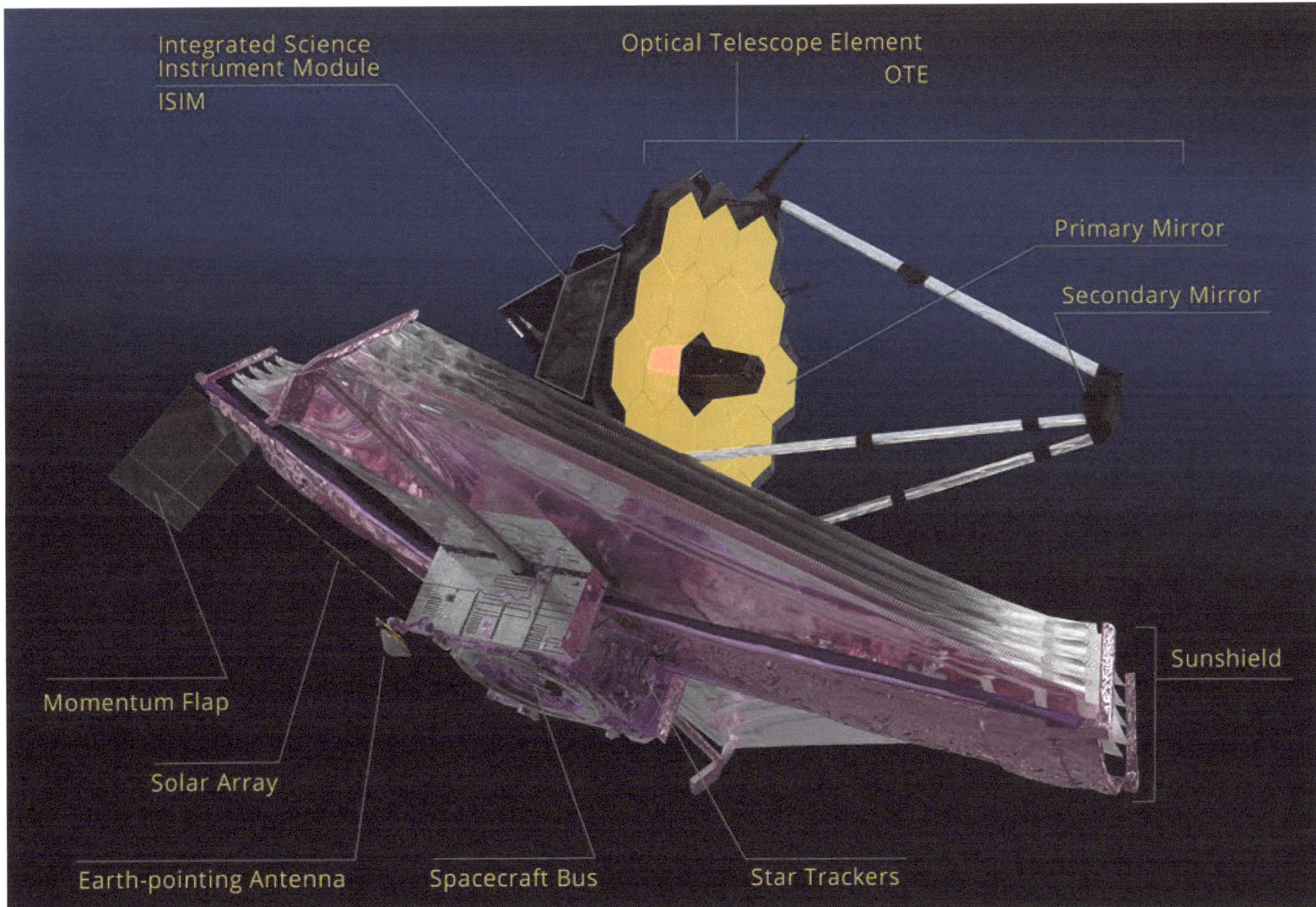

JWST Observatory Elements and Regions

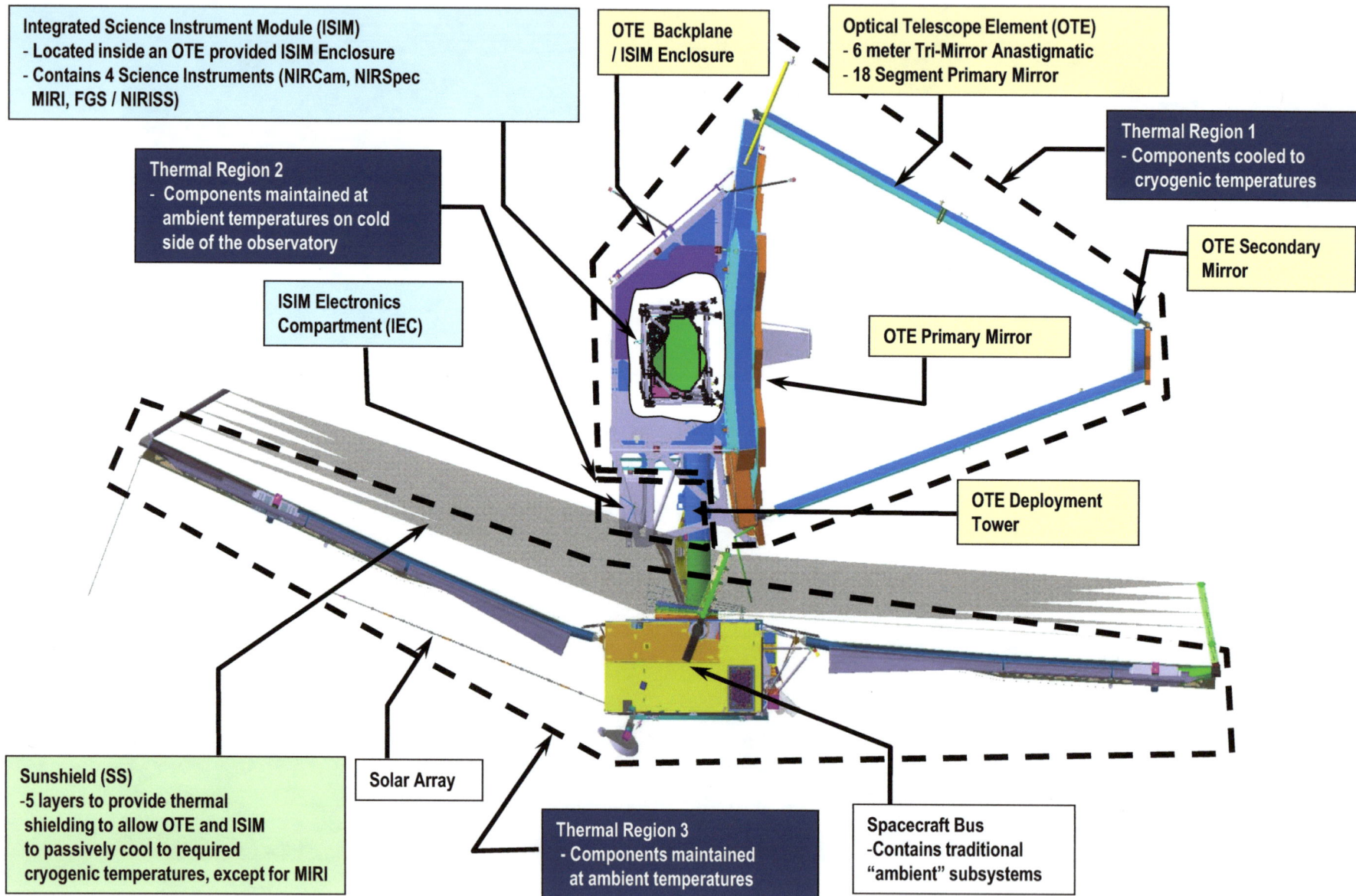

Integrated Science Instrument Module (ISIM)
- Located inside an OTE provided ISIM Enclosure
- Contains 4 Science Instruments (NIRCam, NIRSpec MIRI, FGS / NIRISS)

OTE Backplane / ISIM Enclosure

Optical Telescope Element (OTE)
- 6 meter Tri-Mirror Anastigmatic
- 18 Segment Primary Mirror

Thermal Region 1
- Components cooled to cryogenic temperatures

Thermal Region 2
- Components maintained at ambient temperatures on cold side of the observatory

ISIM Electronics Compartment (IEC)

OTE Secondary Mirror

OTE Primary Mirror

OTE Deployment Tower

Sunshield (SS)
-5 layers to provide thermal shielding to allow OTE and ISIM to passively cool to required cryogenic temperatures, except for MIRI

Solar Array

Thermal Region 3
- Components maintained at ambient temperatures

Spacecraft Bus
-Contains traditional "ambient" subsystems

JWST vs. Hubble primary mirrors

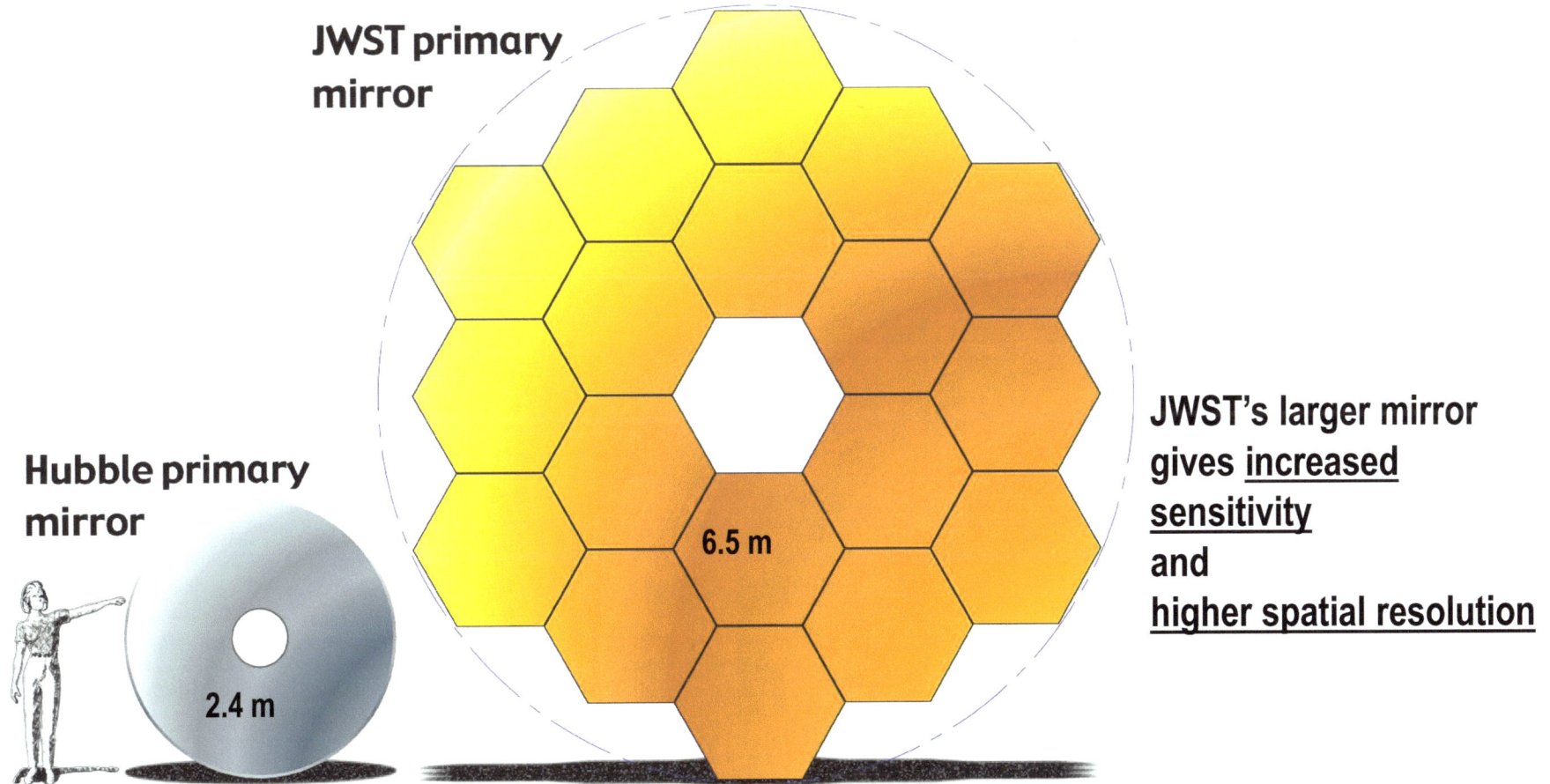

JWST primary mirror

Hubble primary mirror

2.4 m

6.5 m

JWST's larger mirror gives increased sensitivity and higher spatial resolution

Optical Telescope Element (OTE)

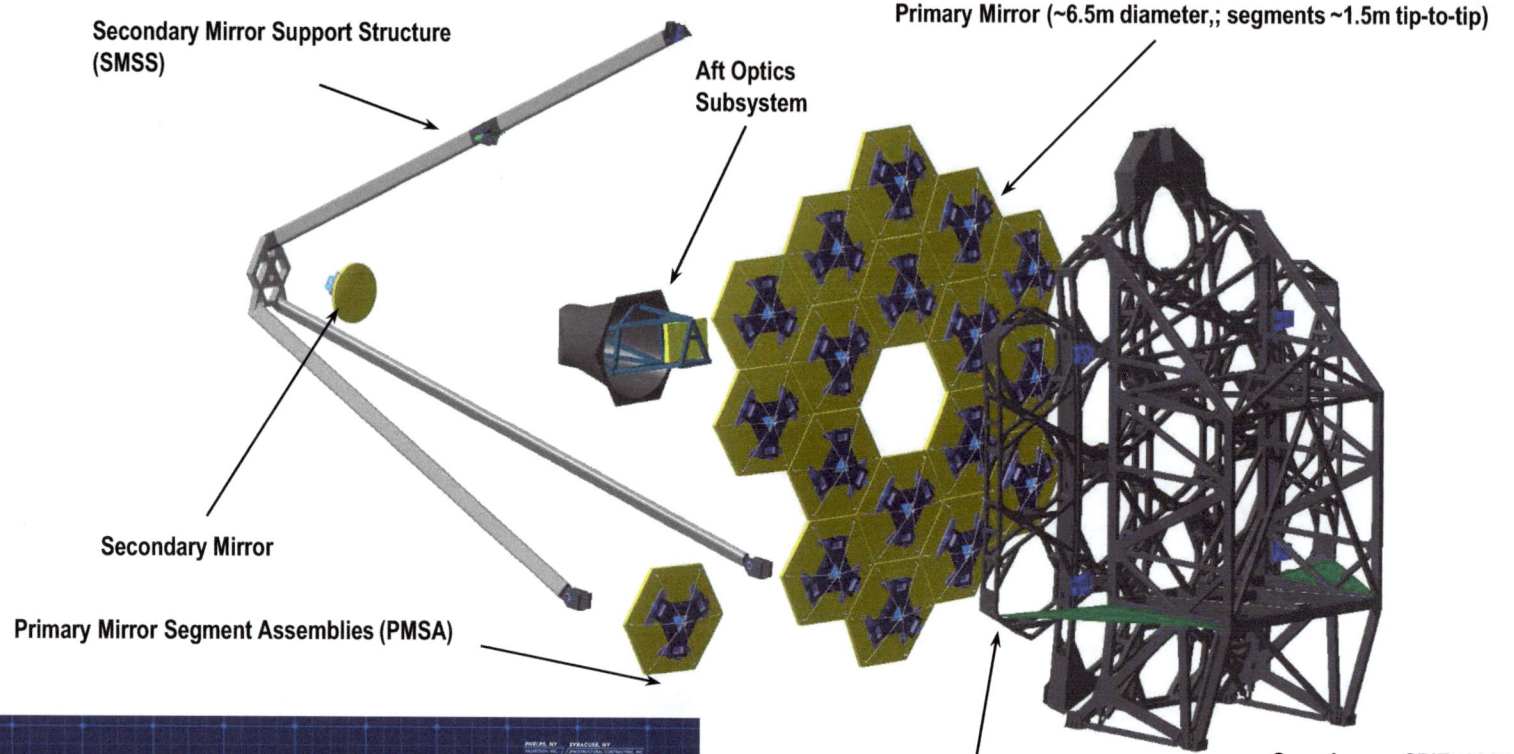

Secondary Mirror Support Structure (SMSS)

Aft Optics Subsystem

Primary Mirror (~6.5m diameter,; segments ~1.5m tip-to-tip)

Secondary Mirror

Primary Mirror Segment Assemblies (PMSA)

Greenhouse, SPIE, 2015

Primary Mirror Backplane Assembly (PMBA) and Backplane Support Frame (BSF)

- Composite tube frame construction
- Two deployable Wings

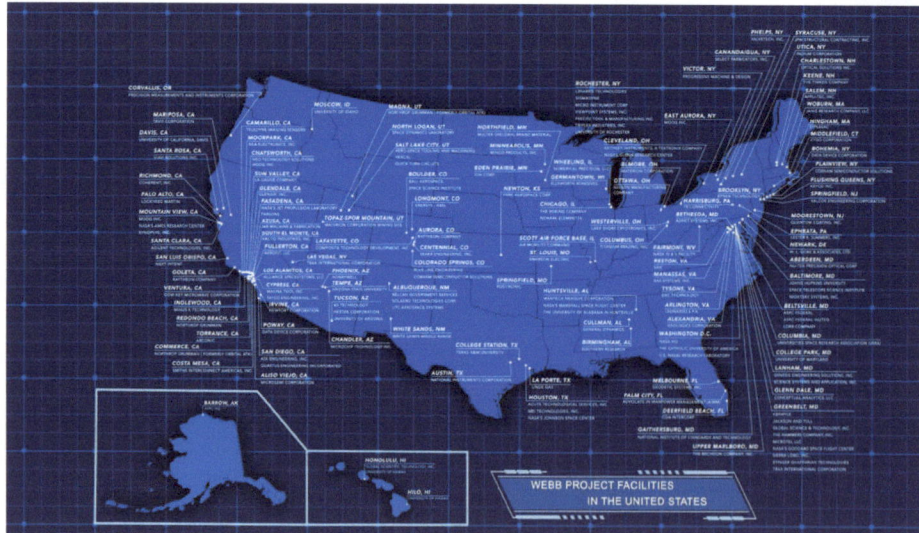

WEBB PROJECT FACILITIES IN THE UNITED STATES

NASA

Videos --- let's see if we can get these to work

- **Some things to note:**
  - The Earth and Sun are always on the spacecraft (warm) side of the sunshield, like a beach umbrella
  - JWST is too big for a rocket nose cone (fairing), so it was launched folded up, then parts deployed as it headed out to L2, like a Transformer toy

- **Deployment video: https://www.youtube.com/watch?v=RzGLKQ7_KZQ**
- **Launch: https://www.youtube.com/watch?v=v6ihVeEoUdo**
- **Video gallery:**
  - https://webb.nasa.gov/content/multimedia/videos.html

Primary mirror segments tested at NASA Marshall Space Flight Center

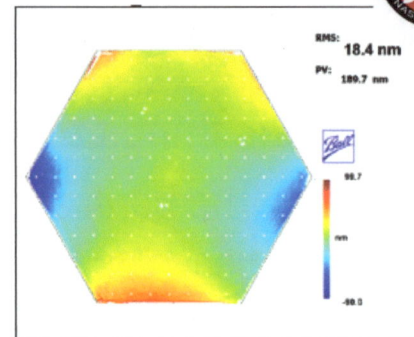

A2

RMS: 18.4 nm
PV: 189.7 nm

Ball Aerospace & Technology Corp.
Secondary mirror testing

L. Feinberg, AAS, Jan 2010

OTE build-up at NASA Goddard (L3 Harris & GSFC)

- **Mirror metrology**
- **Integration and alignment**

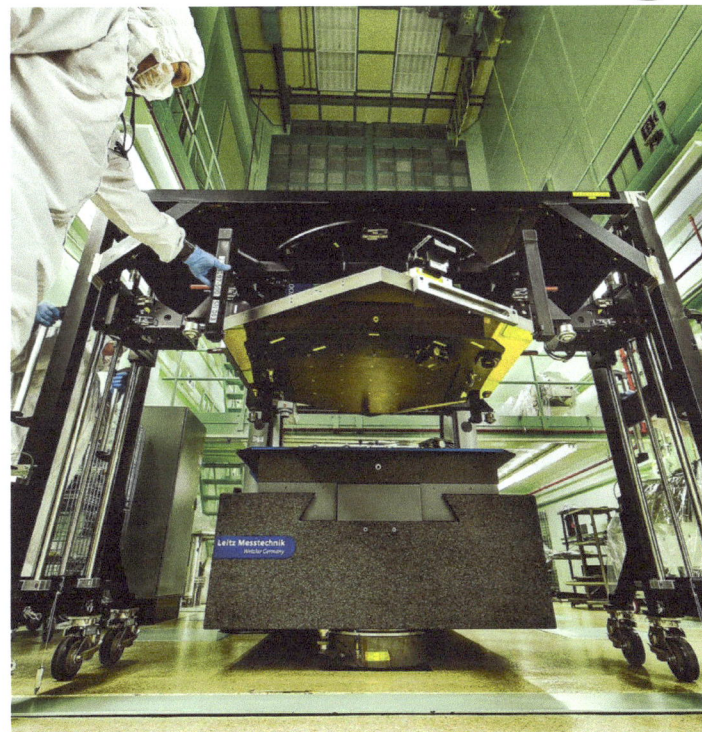

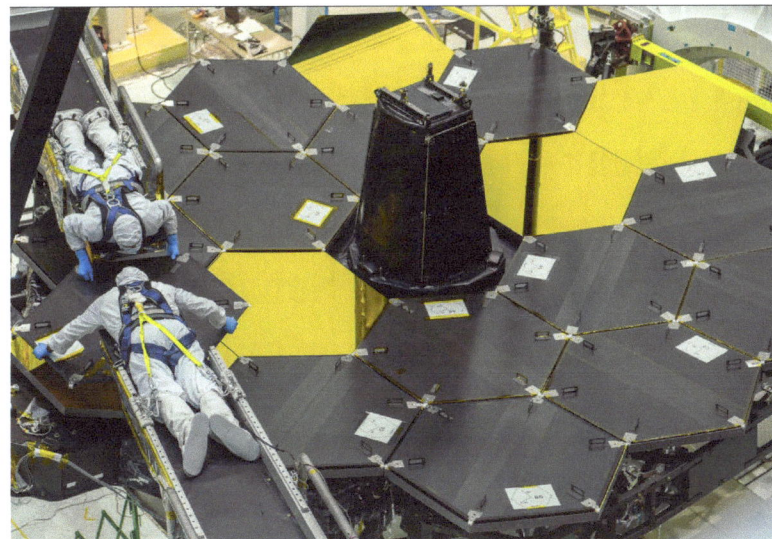

NASA

NASA

NASA

Integrated Science Instrument Module (ISIM)

Integrated Science Instrument Module (ISIM)

- **Housed behind the primary mirrors**
- **Aligned to exit pupil and focal surface of telescope**

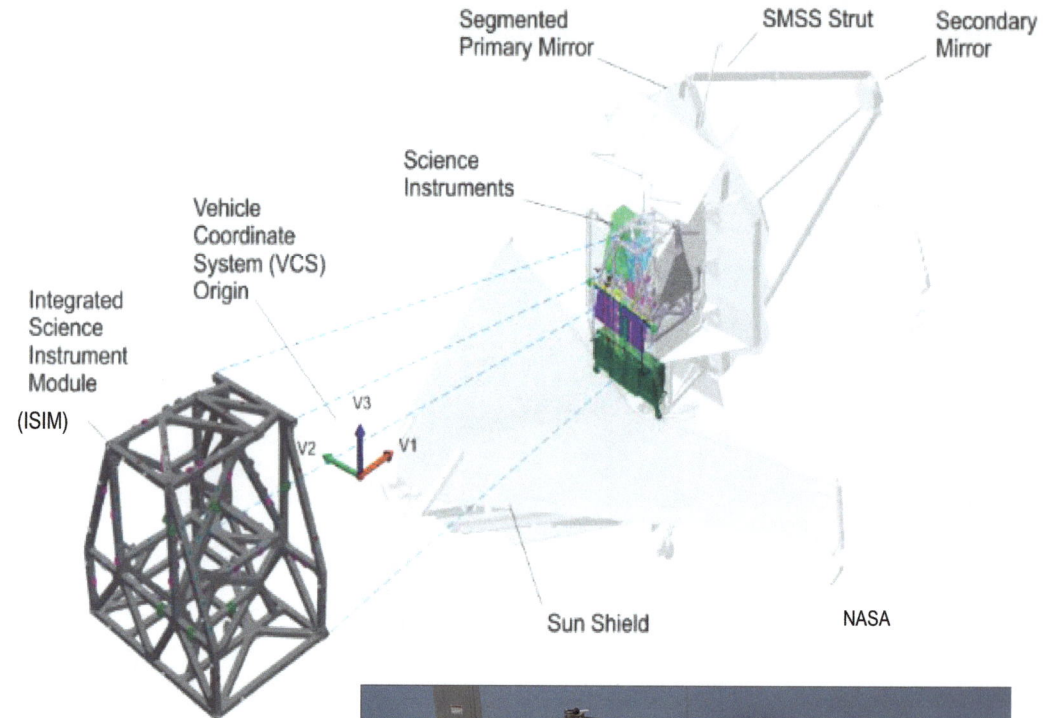

Four instruments on ISIM

- **Near-Infrared Camera (NIRCam)**
- **Near-Infrared Spectrograph (NIRSpec)**
- **Mid-Infrared Instrument (MIRI)**
- **Fine Guidance Sensor (FGS) and Near-InfraRed Imager and Slitless Spectrograph (NIRISS)**

ISIM Structure (ATK, Utah and GSFC)

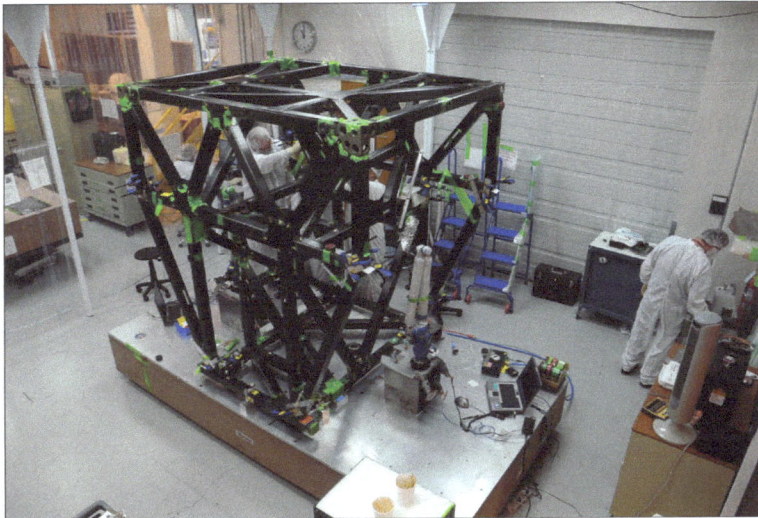

Structure bonding at ATK (Utah)   NASA

NASA

Structure on kinematic mounts (from L3 Harris)

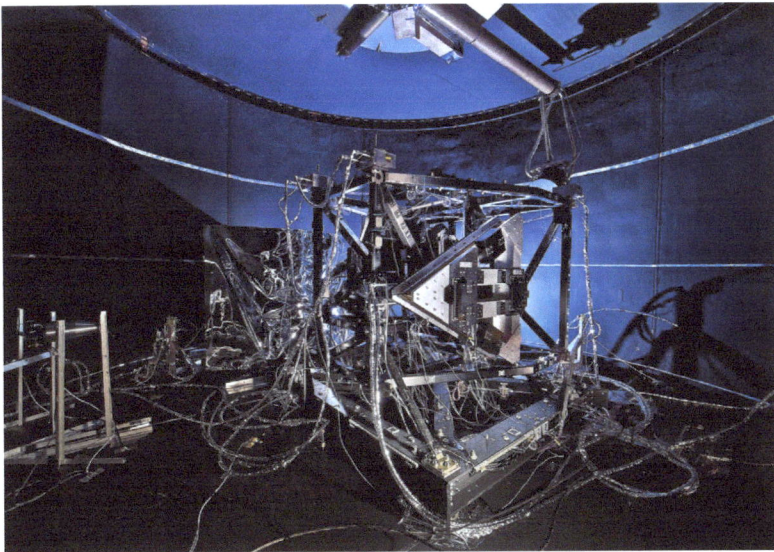

NASA

Structure cryogenic metrology testing (photogrammetry)

NASA

Integration of science instrument optical assemblies

Characterizing & customizing the ISIM Structure

NASA

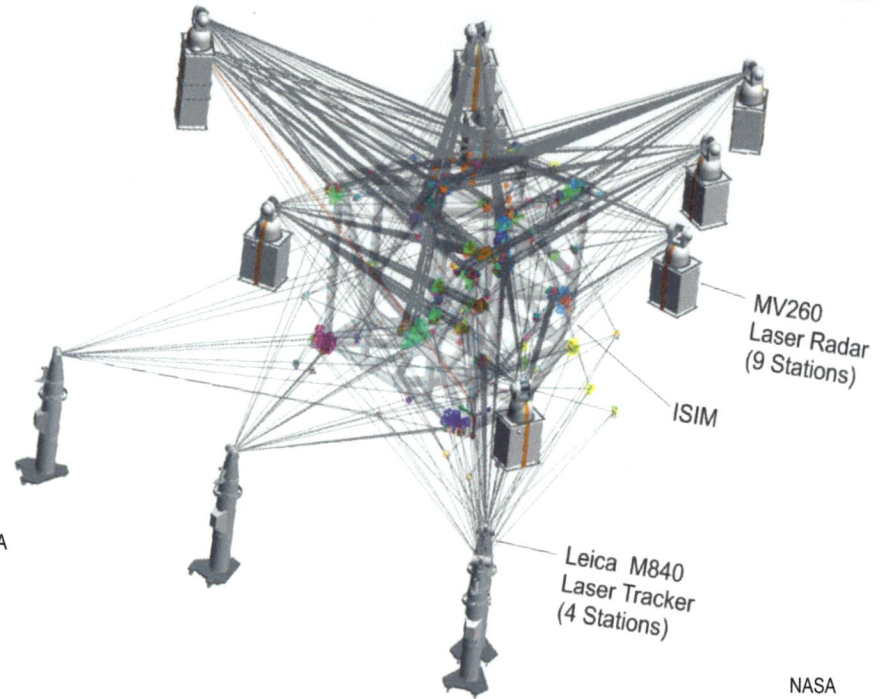

MV260
Laser Radar
(9 Stations)

ISIM

Leica M840
Laser Tracker
(4 Stations)

NASA

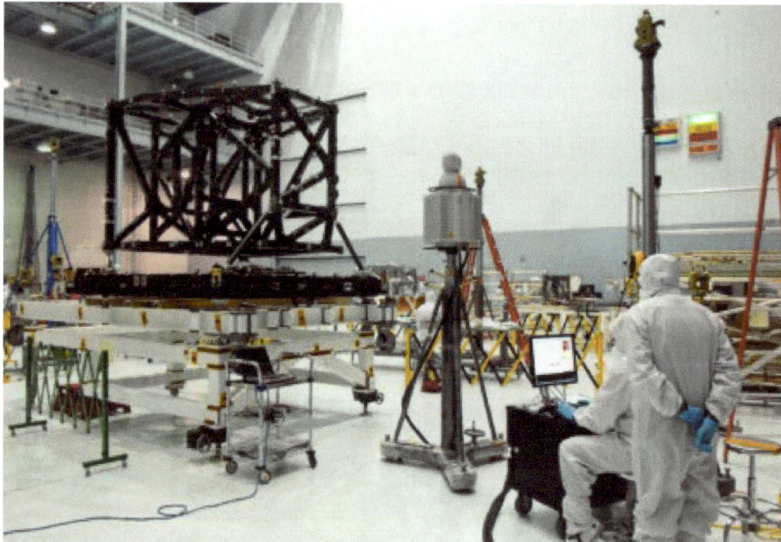

NASA

NASA

23

Science instruments

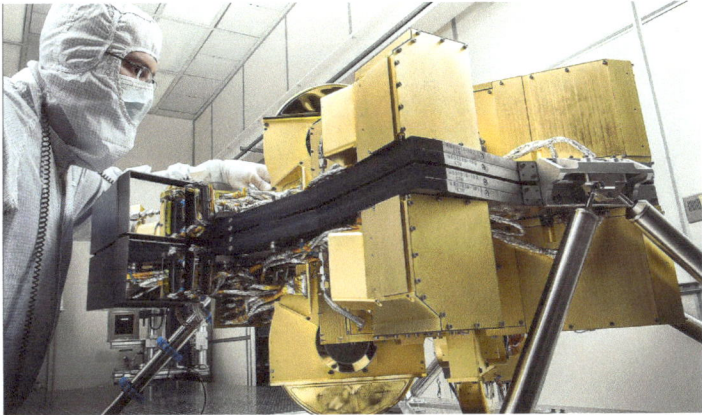

Near-Infrared Camera (NIRCam)
University of Arizona
Lockheed Martin, Palo Alto

Near-Infrared Spectrograph (NIRSpec)
European Space Agency (ESA)
NASA Goddard Space Flight Center

Fine Guidance Sensor (FGS)
Near-Infrared Imager and Slitless Spectrograph (NIRISS)
Canadian Space Agency

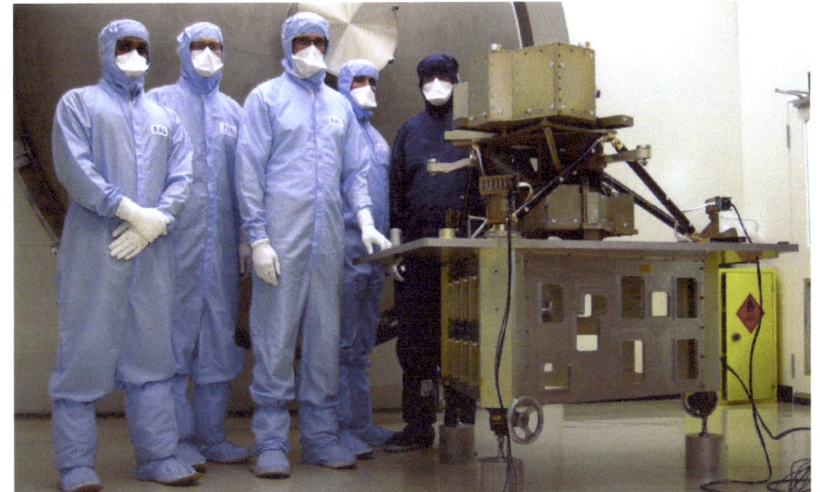

Mid-Infrared Instrument (MIRI)
European Consortium, ESA, NASA JPL

24

Detector and MEMS micro-shutter technology

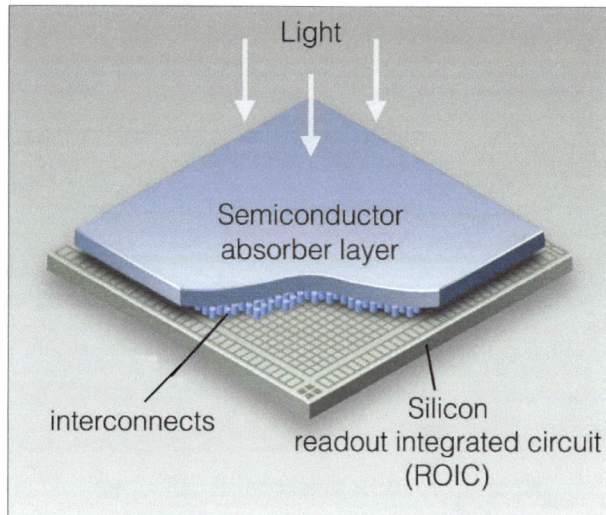

Light

Semiconductor absorber layer

interconnects

Silicon readout integrated circuit (ROIC)

NASA

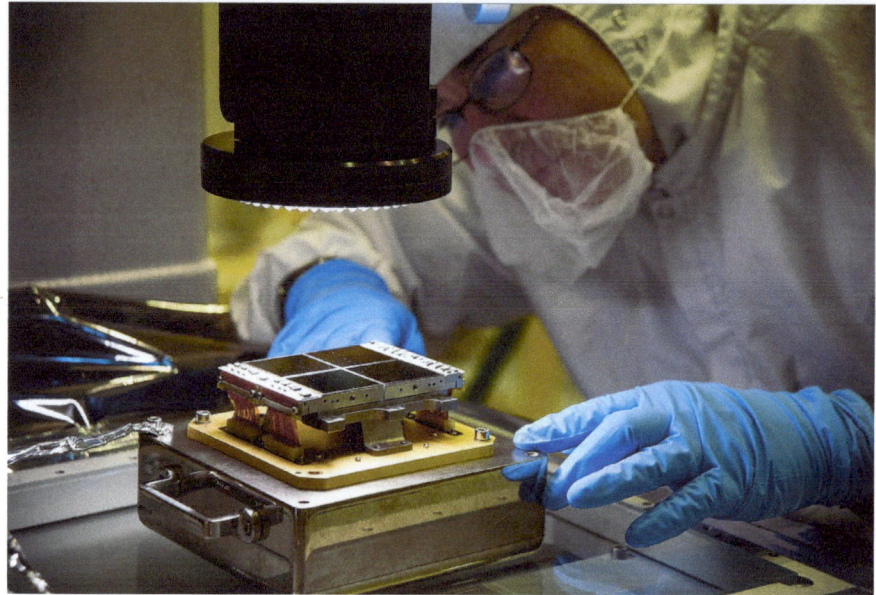

NASA

NASA

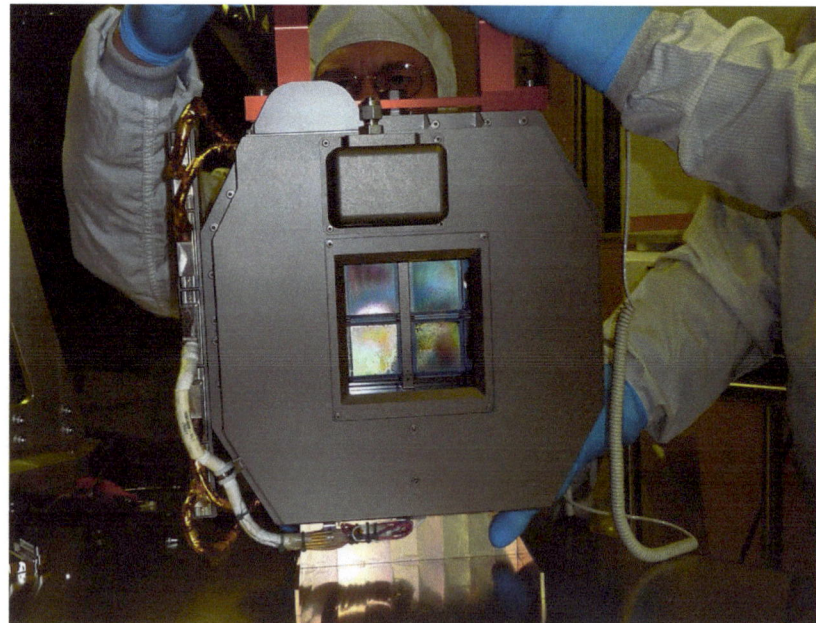

NASA

2017: JWST at NASA Goddard Space Flight Center

NASA

NASA

NASA

NASA

Cryogenic performance testing

NASA Johnson Space Center

NASA Goddard Space Flight Center

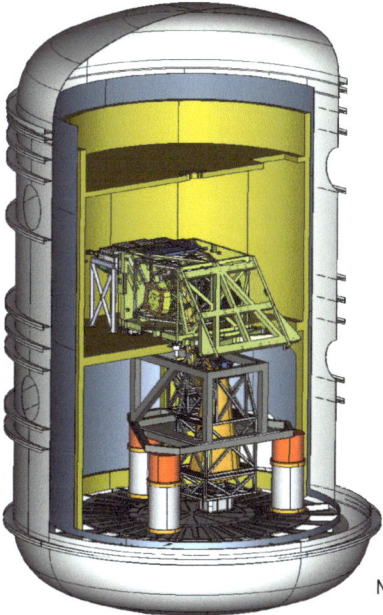

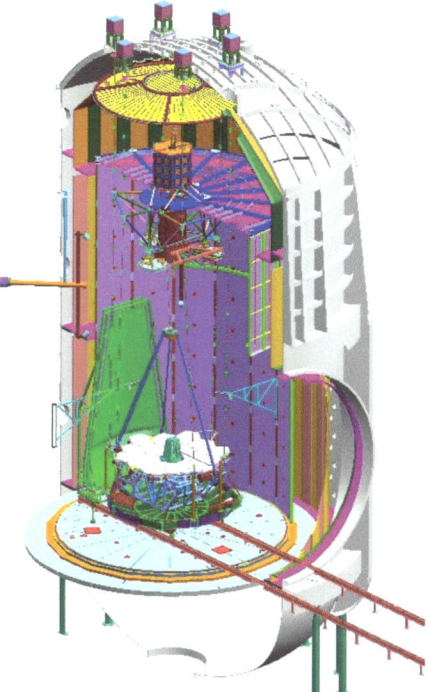

NASA

NASA

NASA

28

2017: JWST optical test: Historic Chamber A

The Apollo era

NASA

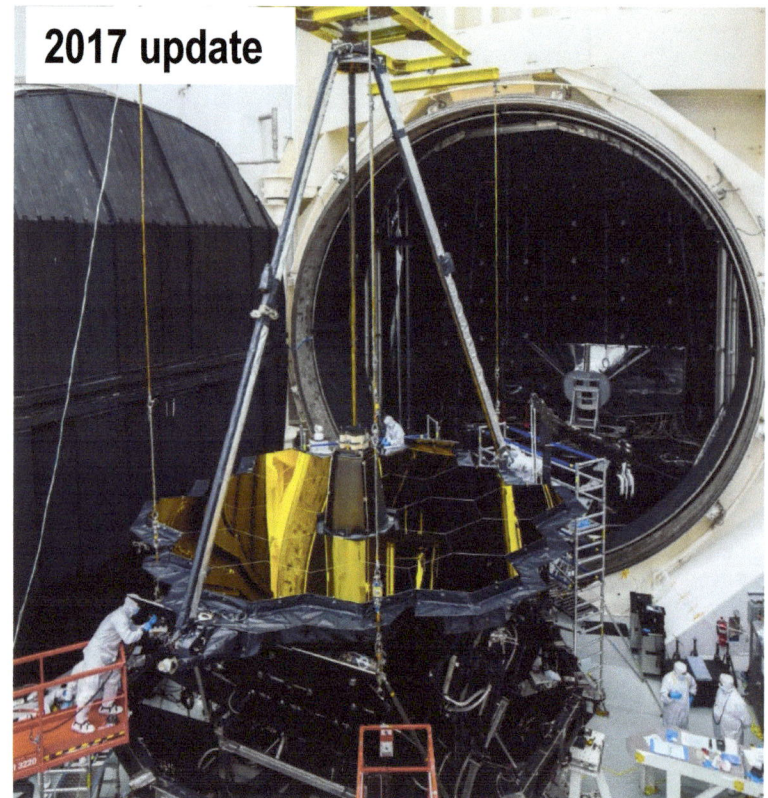

2017 update

NASA

National Historic Landmark at NASA Johnson, outfitted for JWST

2019-20: JWST integration to spacecraft bus

**Northrop Grumman
Los Angeles, CA**

NASA

2021: Final Deployment Testing

**Northrop Grumman
Los Angeles, CA**

NASA

Cover removed from Aft Optics

NASA

2021 (September): JWST Travels to ESA facility

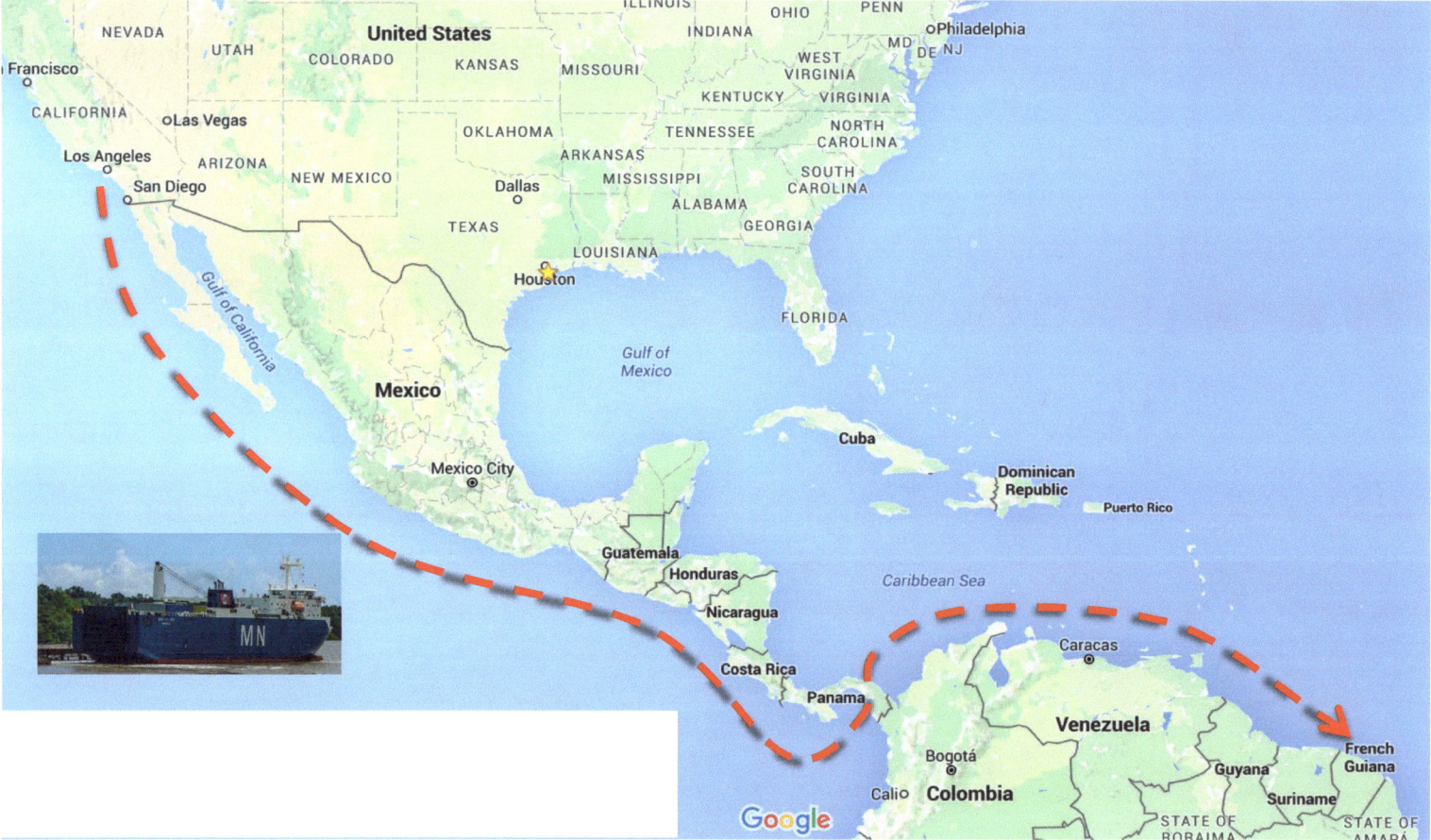

NASA

October 2021: **JWST** at the Launch Site

Oct 12, 2021
RELEASE 21-132

NASA's Webb Space Telescope Arrives in French Guiana After Sea Voyage

After the custom-built shipping container carrying Webb is unloaded from the MN Colibri, Webb w
Credits: NASA/Chris Gunn

Credit: ESA

NASA's James Webb Space Telescope successfully arrived in French Guiana Tuesday, after a 16-day journey at sea. The 1,500-mile voyage took Webb from California through the Panama Canal to Port de Pariacabo on the Kourou River in French Guiana, on the northeastern coast of South America.

The world's largest and most complex space science observatory will now be driven to its launch site, Europe's Spaceport in Kourou, where it will begin two months of operational preparations before its launch on an Ariane 5 rocket, scheduled for Dec. 18.

34

Launch, 25 Dec 2021

NASA

NASA

NASA

ESA

ESA

NASA

NASA

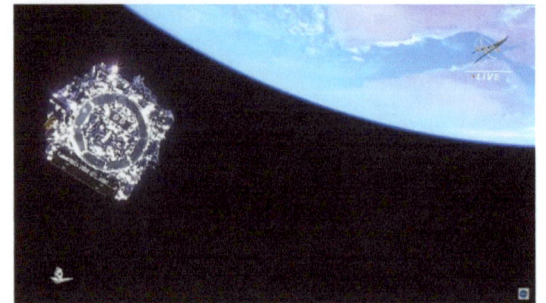

NASA

On-orbit commissioning (NASA JWST blog)

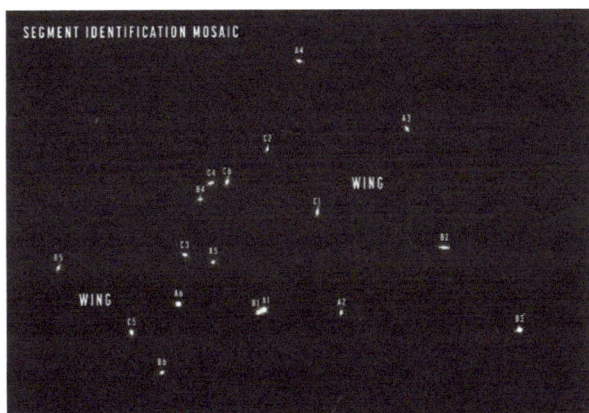

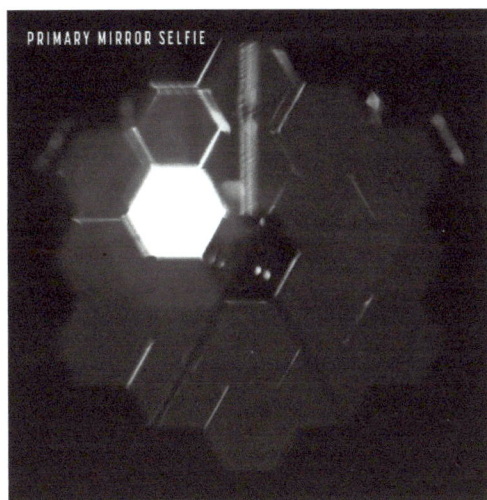

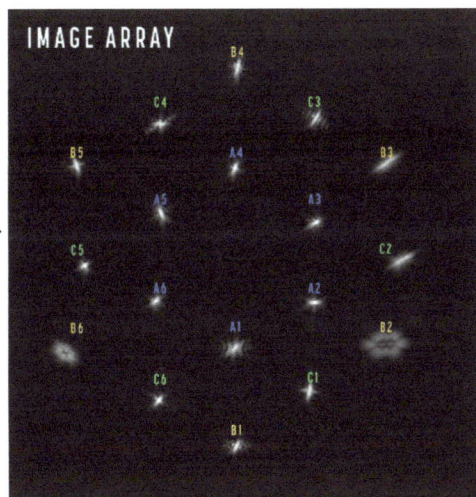

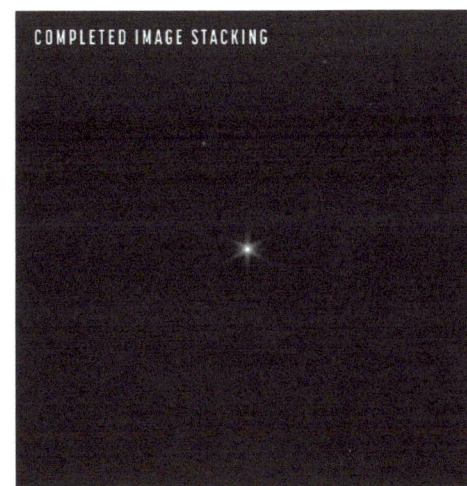

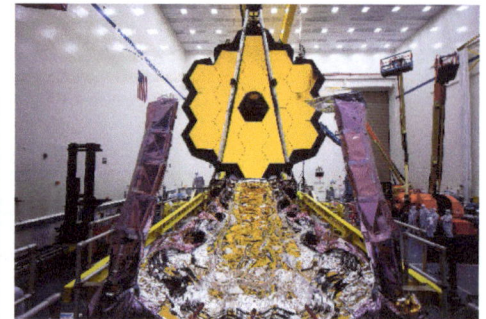

Timescale to build a large space telescope…

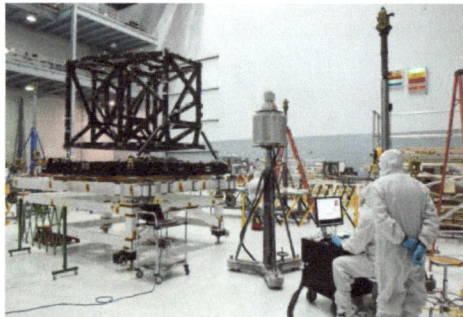

NASA

NASA

NASA

NASA

time

More about JWST

- www.jwst.nasa.gov
- http://www.jwst.nasa.gov/science.html
- https://blogs.nasa.gov/webb/
- https://www.flickr.com/photos/nasawebbtelescope/

Questions?

www.ingramcontent.com/pod-product-compliance
Lightning Source LLC
Chambersburg PA
CBHW041448200326

41518CB00004B/190